Acknowledgements

We would like to express our gratitude to Dr Tobias Gabel and Mr Jeremy Groome for helping us with English proofreading and improving the readability.

We would also like to thank Ms Emilija Kornakova for the visualization and graphical design of the book.

We thank the American Association for the Advancement of Science for permission to reproduce Figure 1. (Wilfert et al, 2016)

We thank The Nature Conservation Programme in North Macedonia (NCP) Bregalnica for permission to use the videos produced during the project: "Alternative methods for varroa control in honey bee colonies" (Uzunov & Büchler, 2020).

Finally, we thank all beekeepers, bee experts and scientists around the globe for their contribution in the development, improvement and propagation of the methods in the beekeeping community.

First published in Great Britain in 2023 by Apoidea Press, an imprint of Bee Publishing Limited, Woodside Cottage, Dragons Lane, Shipley, West Sussex, RH13 8GD, UK.

www.apoideapress.com

Copyright © 2023 by Aleksandar Uzunov, Martin Gabel, Ralph Büchler.

Aleksandar Uzunov, Martin Gabel and Ralph Büchler have asserted the authors' rights under the Copyright, Designs and Patents Act 1988 to be identified as authors of this work.

All rights reserved. No part of this work may be reproduced or utilised in any form or by any means, electronic or mechanical, including photocopying, recording or by any information storage and retrieval system without the prior written permission of the publisher.

ISBN: 978-1-916612-00-6

Layout and design by: Emilija Kornakova

Printed by: Lightning Source.

CONTENTS

1. Introduction **5**
2. Methods for summer brood interruption **13**
   - Brood removal **15**
   - Queen caging **26**
   - Trapping comb **36**
3. Factors affecting applicability **49**
4. Additional aspects and tips **57**
5. Frequently Asked Questions & practical experience **67**
6. Checklist **73**
7. Further reading **77**
8. About the authors **80**

1 INTRODUCTION

INTRODUCTION

The destructive nature of the parasitic varroa mite (*Varroa destructor*) on the European honey bee (*Apis mellifera*) has been well studied and documented. Following the mite's "conquest" of Europe (Figure 1), from its original host (*Apis cerana*) in south-east Asia in the middle of the last century, the attention of beekeepers and scientists alike was initially directed towards using chemical-based treatments for keeping colony mite infestation under control. Significant efforts and investments were made to develop and use synthetic chemicals since their miticide effects were rapid and at first effective. Unfortunately, the unsystematic and excessive use of these drugs led to the equally rapid occurrence and spread of mite resistance in some areas, leading beekeepers into a whirlwind of drug overuse.

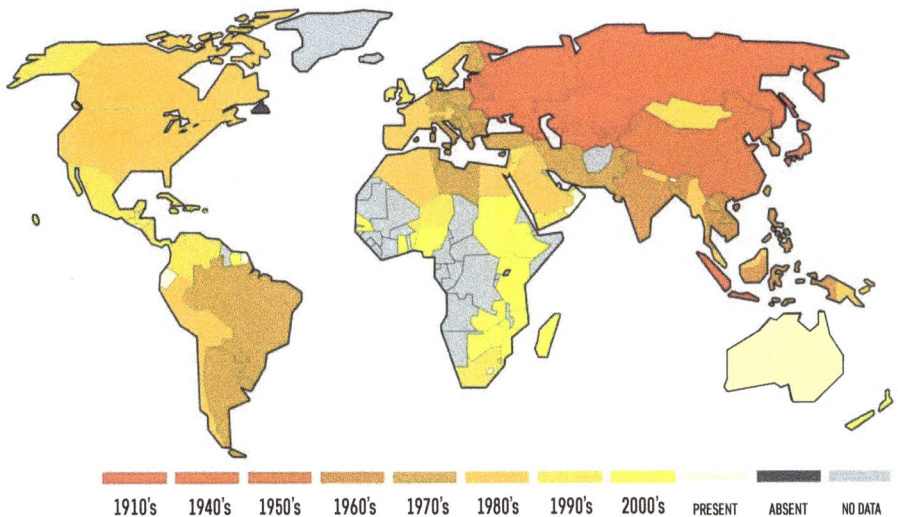

Figure 1. The global spread of the varroa mite (*V. destructor*) from its native Asian distribution range, which has led to its current role as one of the biggest threats to apiculture worldwide (modified from Wilfert *et al.*, 2016; reprinted with permission of AAAS).

Consequently, beekeeping nowadays faces additional problems such as a significant ineffectiveness of chemical-based treatments and a reduction in the safety of hive products due to residues in honey and beeswax. Chemical residues, especially, stand in sharp contrast to consumers' common understanding of hive products as being natural and uncontaminated. Such side effects of conventional varroa treatments underline how outdated those once common methods are, in the light of today's better knowledge, more demanding market requirements and raised consumer awareness.

In parallel to this chemical-based approach, an alternative also emerged from enthusiasts whose beekeeping philosophy was more in tune with the honey bee colony's biology, environmental protection, and food safety. Using a better understanding of the development dynamics of honey bees and varroa mites, their respective behaviours and interactions with each other, these beekeepers and scientists developed treatment methods entirely without or with only limited use of chemicals. The peculiarly synchronised life cycles of the mite and the honey bee colony were recognised not just as an opportunity, but as the key starting point for controlling the parasite. In fact, both phases (Figure 2) of the mite's life cycle, the reproductive phase (in the brood) and the non-reproductive or "phoretic" phase (on the adult bees), can be utilised for the control of colony infestation.

PHORETIC PHASE

The frequently used term "phoretic phase" as a description for the part of the mite life cycle spent on adult bees is strictly speaking incorrect with regard to the varroa mite. In biology, phoresis is a form of commensalism and thus defined as an interaction in which none of the parties involved is harmed (Matthes, 1978; White, Morran, & de Roode, 2017). Since adult honey bees are in fact parasitized and thereby harmed by varroa mites (Ramsey et al. 2019), theirs is not a phoretic relationship in the strict sense. For reading convenience, however, we will keep the commonly accepted term "phoretic".

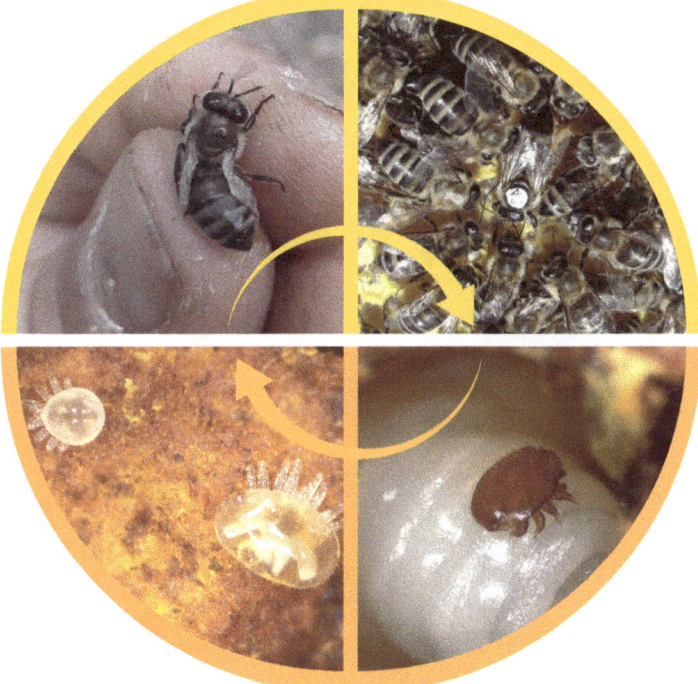

Approx. 7 days
Free living phase on adult bees

Reproductive phase in capped brood cells
12 or 14 days

Figure 2. The life cycle of the varroa mite.

Swarming, and less commonly absconding, are well-known phenomena in a honey bee colony's life cycle. In both cases, the colony is mostly separated into bees and brood, followed by an interruption of brood-rearing (Figure 3). This is not only known to spontaneously reduce the colony's varroa population and other pathogens but has also been shown to reduce the possibility of reproduction for the remaining mites (Figure 4). Besides having reproductive and migratory functions, swarming with its beneficial side effects, acts as a natural remedy mechanism that plays a key role in the survival of unmanaged honey bee colonies.

Figure 3. A problem shared is a problem halved: During swarming, the varroa population of the hive is split up along with the bees. In addition, both parts – the swarm and the swarmed colony – face a subsequent brood break. Both factors inhibit varroa reproduction, and thus support hive health.

On the other hand, this sharply contrasts with widespread beekeeping techniques, which commonly aim for high honey harvests by swarm prevention and supported brood rearing. However, continuous brood rearing throughout the season inevitably leads to an exponential growth of the mite population and a critical brood infestation level until the time of winter bee development (Figure 5). Thus, an urgent need was raised to integrate the inherent resistance mechanisms of the honey bee colony into modern, applicable colony management.

By learning from the biology of the host (bee) and the parasite (mite) while following the needs of beekeepers for adequate colony management techniques, beekeepers and scientists jointly developed various biotechnical methods for varroa control, and thereby integrated the honey bee's natural survival mechanisms into modern beekeeping.

Hence, effective mite abatement can be achieved with a set of so-called "summer brood interruption" methods that produce either a real interruption or at least a reduction in brood rearing, and are particularly attractive to beekeepers who aim to integrate near-natural methods into their varroa control practice. For example, by removing reproducing mites from the brood cells and then exposing the mites remaining on the adult bees to a comparatively low dosage of highly targeted organic acaricides (or even by trapping them in a subsequent brood comb without any use of chemicals), beekeepers are able to remove a significant portion of a colony's mites whenever needed – even during the nectar flow.

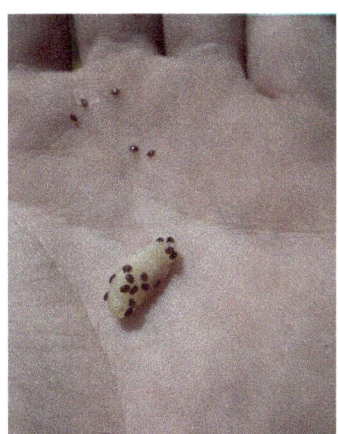

Figure 4. Even though varroa mites harm both brood and adult bees, the main damage is inside the brood cells, which are also crucial for mite reproduction and population growth.

These methods can be combined with the use of natural compounds such as organic acids to increase the efficacy in the fight against the mites since the general idea behind this approach is to avoid the use of any synthetic ("hard") acaricides and to reduce the overall usage of chemicals within the hive by supporting their effectiveness with an adapted colony management. Brood interruption methods are thus a valid and sustainable varroa control approach, and are of particular importance in many common colony management concepts that, in order to maximise the honey harvest, usually avoid swarming and brood breaks early in the season.

However, such methods are especially suitable for honey bee populations from regions with an extended brood-rearing and reproductive season, as colonies have sufficient time in summer and autumn to "recover" after their application.

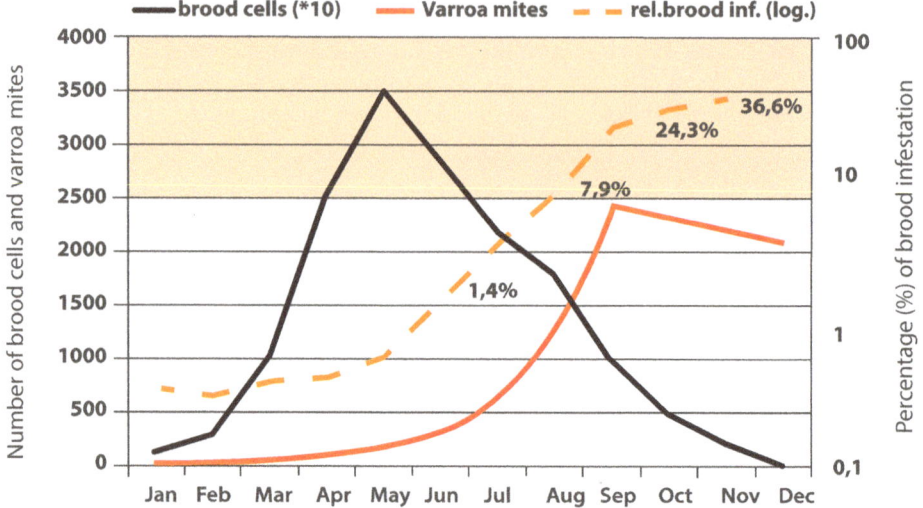

Figure 5. Brood cells and mite population over the course of one year.

Considering the evidence of climate change, and global warming in particular, we expect that both winter and summer honey bee brood-rearing seasonal patterns will be affected (Figure 6). Under such conditions, brood interruption methods may become increasingly important as a practical beekeeping strategy against the varroa mite.

We see significant advantages in practising this approach, particularly in small-scale and hobby beekeeping, but if the methods are adjusted to suit intensive management practices, they can also be incorporated into commercial beekeeping operations. Moreover, they are particularly well suited for organic beekeeping since their utilisation of natural mechanisms reduces the use of chemicals.

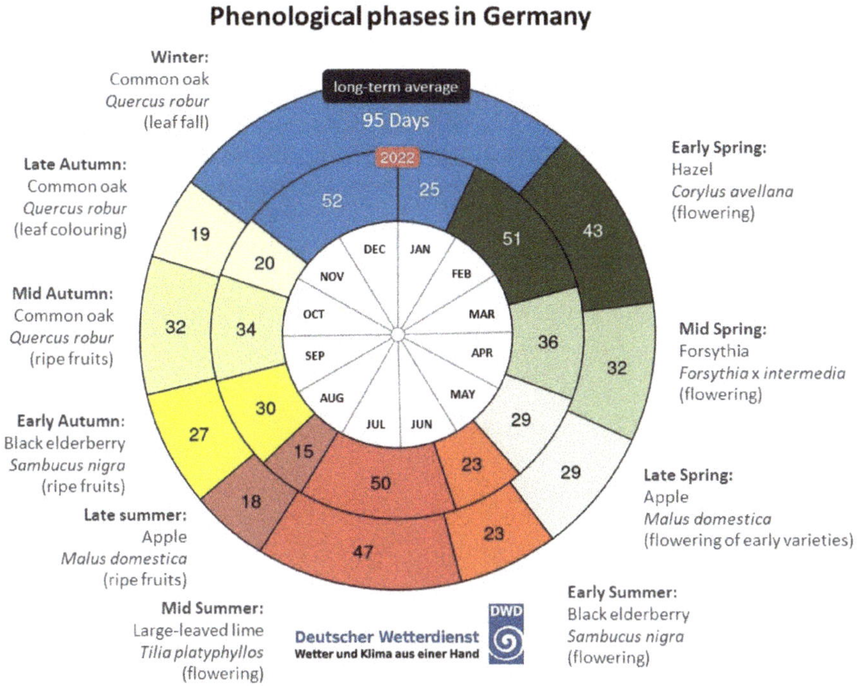

Phenological phases in Germany

Figure 6. Average beginning and duration [Days] of main seasons are shown for the year 2022 (inner circle) and the long term average. Modified after Deutscher Wetterdienst.

This book aims to provide a detailed description based on the understanding and use of biological mechanisms, and a step-by-step manual for the most prominent biotechnical methods of varroa control. These systematic guidelines are the outcome of our longstanding beekeeping and research experience with brood interruption methods in central and south-eastern Europe, while having also benefitted from the knowledge gathered by many other beekeepers and researchers throughout Europe. Thus, we also incorporate other authors' main findings and provide many practical tips to reduce the risks of a queen or colony failing during the treatment proposed below.

Finally, we hope that this material will fill a gap in the beekeeping literature about the application of biotechnical methods for varroa control, thereby leading to a more sustainable management concept that is in tune with honey bee biology, consumers' requirements, and an increasing public awareness of environmental concerns.

2
METHODS FOR SUMMER BROOD INTERRUPTION

METHODS FOR SUMMER BROOD INTERRUPTION

To combat and control the varroa mite population in the honey bee colony, any of three principal brood reduction or brood interruption methods may be used:

- **brood removal**, when either part or all of an infested colony's brood is removed, along with a significant portion of the mites present, and the remaining mites are exposed for mitigation;
- **queen caging**, in which full brood interruption is achieved by confining the queen in a cage without the possibility of laying any more eggs – a situation in which all the mites in the colony become exposed for treatment on the adult bees;
- the **trapping comb** technique, a combination of both the aforementioned methods which makes possible the removal of the mites trapped alongside the brood reared in particular combs.

No single one of these methods is a perfect fit everywhere, and beekeepers should always consider the respective complexities, time and labour requirements when choosing which method to use. Moreover, the possibilities for combining or adjusting these methods to different beekeeping conditions and needs are virtually endless.

To get a first idea of which method could fit your beekeeping interests, Figure 7 provides an overview of the main differences between the three options in terms of hive management.

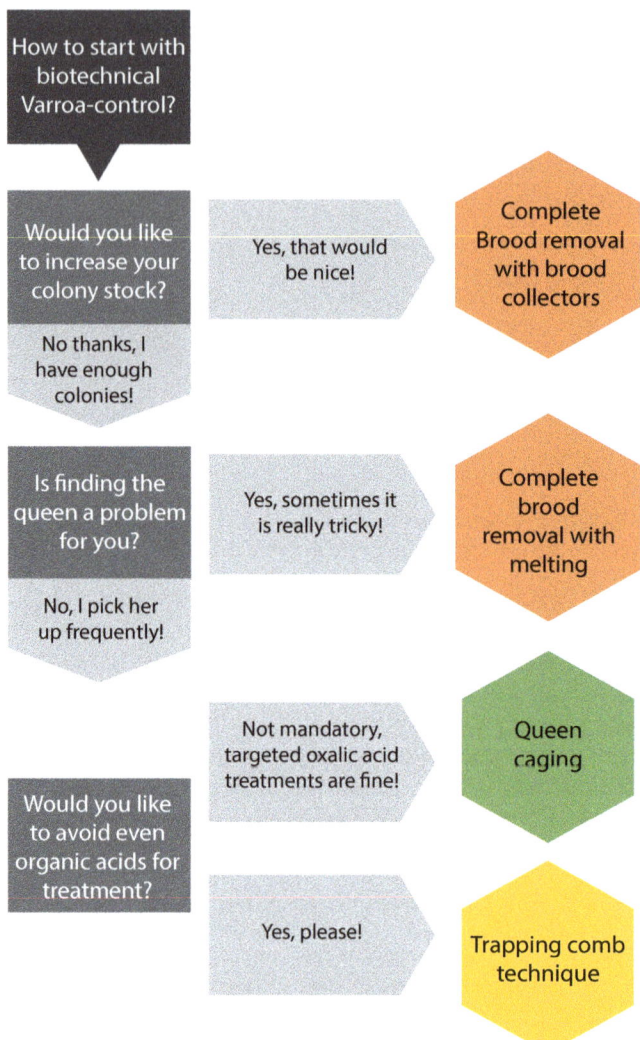

Figure 7. Decision tree for identifying the method of brood interruption best suited to your beekeeping practice.

To guide you to the most suitable method for your hive management and beekeeping style, in the following section the three methods of brood interruption are described in detail as well as in practical, step-by-step manuals.

SUMMER BROOD INTERRUPTION FOR VITAL HONEY BEE COLONIES

BROOD REMOVAL

... to watch the video for method application

Courtesy of the Nature Conservation Programme in North Macedonia – NCP[1]

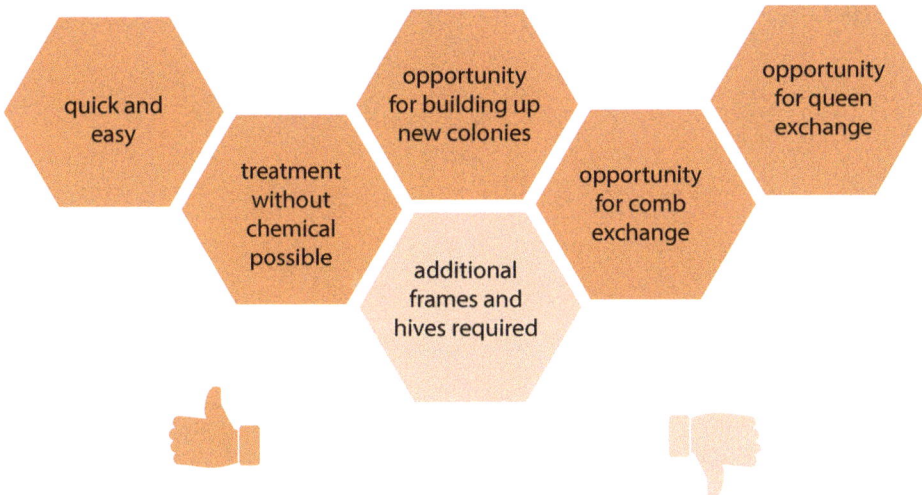

At a glance

For the brood removal method, no "special" tools are needed; you will work with basic beekeeping equipment (e.g. frames and hive boxes; Figure 8). However, you might need extra hive boxes, lids, and bottom boards if you plan to use the removed brood for building nucs (brood collectors). Basically, the mites inside the brood cells are removed alongside all brood combs before treating the remaining mites on the bees with oxalic acid or trapping them in another brood comb. It is helpful to find and catch the queen, although it is not necessary for this method to work.

[1] "Alternative methods for varroa control in honey bee colonies", Uzunov and Büchler, 2020, Nature Conservation Programme in North Macedonia – NCP.

Figure 8. Equipment needed for the brood removal method. For the basic method, you only need frames with foundation or empty combs.

Basic idea

The removal of all of a colony's brood might be imagined to be a massive disruption of that colony's natural brood cycle, but in fact, it is inspired by the fascinating behavioural patterns exhibited by feral honey bee colonies. For instance, during a natural swarm event, the colony is separated into two components: the highly infested brood with a minor portion of the adult bees and the slightly infested majority of adult bees (the swarm) with the "old" queen. This split of the varroa population brought about by this event, as well as the subsequent brood interruptions occurring in both parts of the colony, are key mechanisms in naturally varroa-surviving colonies.

On the other hand, this situation, with the infested brood in the hive and the swarm somewhere outside, is quite inconvenient for modern beekeeping, especially when it occurs before or during the main nectar flows (Figure 9). By performing brood removal, beekeepers may tailor these benefits for the colony to the needs of their own apicultural practice: the colony is split after the honey harvest, with both parts remaining in the care of the beekeeper.

Figure 9. Natural swarming can lower a colony's mite burden in various ways, but at the same time may lead to lower honey harvests and the loss of swarms, since climbing treetops is not everybody's preferred leisure activity.

Removing all brood combs with the reproductive mites inside is the first step towards an effective treatment. In the second step, you have to get rid of the mites sitting on the bees. In order to achieve this, you can trick them by presenting a single brood comb with larvae inside the broodless colony (Figure 10). The phoretic mites will then seize their only chance for reproduction, invade the open brood cells (Figure 11) and thereby become trapped in the comb. After capping, the "trapping comb" can easily be removed and destroyed along with the mites "trapped" inside. Alternatively, after the brood removal you can treat the bees with oxalic acid (see page 30) since this organic acaricide is highly effective against the mites in broodless colonies. The broodless period of the colony generated by this technique is also a favourable time for a queen exchange. During this process, the old queen can be replaced by a young one introduced into the broodless colony in a candy- sealed cage.

Figure 10. The ideal trapping comb contains a large amount of young brood (eggs and young larvae). If drone brood is available in a suitable stage, this is even better than worker brood, as the mites prefer to parasitize the latter.

Figure 11. If the "buffet" for mites is scarce, they have to invade the trapping combs provided in order to reproduce.

Use of removed brood combs

The accumulation of brood combs can be both a blessing and a curse of the total brood removal method. Whilst small-scale beekeepers and hobbyists with a limited amount of hive equipment may struggle to find the extra frames and hive boxes needed, the brood removal method enables beginners as well as skilled beekeepers to increase their number of colonies. In any case, they may choose what to do with the removed brood combs depending on their individual needs and possibilities. Basically, there are two options:

1. If, for whatever reason, you don't wish to use the brood, the combs can simply be melted down to harvest the wax, as is common practice for drone combs during the summer (Figure 12, left).

2. If you wish to make use of the brood and have spare equipment, you can gather it in so-called "brood collectors" (Figure 12, right) a special kind of nucleus. If this second option appears to suit your needs, please read the section "Brood collectors" (see page 22) before starting with the brood removal method.

Figure 12. Left: The most straightforward way of using the removed brood combs is to melt them down for harvesting wax, just like old combs or drone brood. Right: If you wish to increase your number of colonies, the brood combs may also be used for brood collectors. As a rule of thumb, brood combs from two colonies will create one brood collector with two boxes.

CREATING STRONG AND HEALTHY COLONIES IN TWO STEPS!

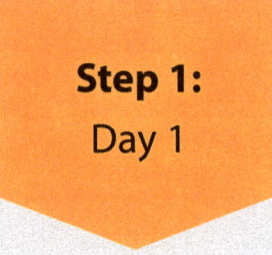

Step 1: Day 1

STEP	DAY
1	1
2a	2
	3
	4
	5
	6
	7
	8
2b	9

* Prepare new combs or frames with foundation to exchange the removed combs as well as empty hive boxes with bottom boards and lids.

* Begin with the lower brood chamber, in case of hives with more boxes, place the upper ones aside (Figures A & B).

* Exchange empty old combs as well as brood combs from the lower brood chamber with nice pollen and food combs from the upper one (Figure C). Shake or brush bees from all old combs and brood combs in the restructured hive (Figures D & E). If you find the queen transfer her gently direct to one of the remaining combs or place her in an introduction cage sealed with candy to stay on the safe side. If needed, add a second brood chamber and fill it with new combs or frames with foundation.

* Remove the retired combs. Keep them bee safe in sealed hive boxes to prevent robbery (Figure F). See page 22 for the usage options of removed brood combs.

* If you would like to use oxalic acid for the next treatment step you're done for today. Continue with Step 2a.

OR

* If you would like to treat the colony without chemical usage choose one brood comb with only young brood (as much as possible) and mark the top bar. Place it as a trapping comb in a central position in the upper box of the new hive (Figure 13). Continue with Step 2b.

Step 2a:
Day 2

* Treat the colony with oxalic acid according to your local legislation (see page 30).

* Have an eye on the colony's food storage. If the honey flow has stopped already and especially if the bees have to build up new combs from foundation you have to feed them preferably with syrup.

OR

Step 2b:
Day 9

* Exchange the now capped trapping comb with a new comb or foundation. Melt the trapping comb.

* Have an eye on the colony's food storage. If the honey flow has stopped already and especially if the bees have to build up new combs from foundation you have to feed them preferably with syrup.

Figure 13. Left: The trapping comb (marked with X) introduced right after the brood is removed from the colony. Right: a mite invading unsealed larvae on a trapping comb.

Brood collectors

Building up nucs from removed brood combs ("brood collector") is another positive feature of the total brood removal method. In most cases, it is a good and easy way of building up new colonies relatively late in the season. As the name "collector" implies, it is possible to collect removed combs and bees from several different colonies. Usually, the brood combs of two treated colonies will find space in one double-boxed brood collector. Nevertheless, one must keep in mind that the main aim of removing brood combs for varroa treatment is the removal of the reproductive mites inside the cells (Figure 14). Therefore, it is essential to check the infestation level of the respective colonies to decide whether using the brood combs for brood collectors makes sense. In the case of highly infested colonies, we can expect heavily parasitised brood cells from which only weakened and damaged bees will hatch. In this case, the best use of the brood combs is to just melt them down.

Figure 14. Varroa mites are confined in the sealed brood cells (photo: Per Kryger).

On the other hand, it would be a pity to waste the brood combs if the colonies which have to be treated are otherwise healthy enough and you would like to increase your number of colonies. In this case, the brood collectors will need some care to develop into fully grown colonies before the onset of winter. Each brood collector will be equipped with 18–19 brood combs (amounting to two hive boxes of, e.g., Langstroth hive type) and one or two combs of honey. The vast number of emerging young bees will make the newly built nucs grow rapidly, while the honey supplies will keep them fed until they start to gather nectar and pollen. The brood cells will need some caretaking nurse bees for their development and warmth, so you need to leave 250–300 bees (approximately enough to cover the area of your hand) per brood comb, which will be sufficient to ensure proper care. Just leave this number of bees on each brood comb while shaking or brushing the others back into their former hive box. Of course, it is essential to ensure that the old queen is not transferred to the new brood collector colony since the now broodless colony would be lost without her! If you can't find her during the brood removal, just shake all of the bees from the brood combs into the former hive. Then add the required amount of worker bees from the honey supers to the brood combs to form your brood collector in the new hive box. This way, you will stay on the safe side since the honey supers will have been separated with a queen excluder and thus should be "queen-free".

The new brood collectors are threatened by four main dangers which beekeepers should avoid:

⚠️ **Varroa.** The brood collectors will start with the burden of mites inside the brood cells. Depending on the infestation level and the kind of requeening adopted (see the next point), you have several options to treat the brood collectors. If they are heavily infested, or you plan to requeen them with a mated queen, treat the nuclei with formic acid, which effectively targets mites inside brood cells. Do this before introducing the new queen, since it may otherwise lead to queen failure. If your brood collectors are only slightly infested and you leave them to raise their own "emergency" queen, once they have become broodless, you can treat them using oxalic acid (see page 30).

⚠️ **Queenlessness.** The brood collectors will need a good queen to prosper. There are various possibilities for requeening the brood collector. Depending on their respective availability, you can introduce a mated or unmated queen, or just allow the bees in the collector to raise an emergency queen from the youngest brood. If you treat them with formic acid, it is very likely that young brood and therefore also young queen larvae will be harmed. Formic acid should consequently only be used if the intention is to introduce a mated or unmated queen after the treatment is finished, to ensure a healthy queen and good acceptance.

⚠️ **Lack of food.** Especially for the first days, the brood collector will have a limited workforce to collect new food. As described on page 22, the brood collectors will start with honey combs to keep them going until sufficient foragers are available. However, be aware that sometimes a considerable amount of honey can be found even on the brood combs. Usually, this will prove sufficient until the bees in the collector are strong enough to collect food, but you will need to keep an eye on the weather and the amount of nectar and food available and feed them if necessary. Otherwise, especially if there is no nectar flow, there is a risk of starvation.

⚠️ **Robbing.** As with the lack of foragers, the new colonies will have few guards to defend themselves from robbing (Figures 15 & 16). Therefore, for brood collectors, you should apply the same safety measures as are needed for "regular" or smaller nucs. The hive entrance should be kept very small (one or two fingers wide), and the colonies should be placed in a separate apiary without full-sized colonies. The bottom board should be covered with a drawer not only to count fallen varroa mites but also to support thermoregulation and prevent robbing.

Figure 15. A weak colony is unable to defend a wide entrance and gets robbed.

Figure 16. A similar-sized colony with a reduced entrance can defend itself better against wasps and robbing honey bees.

SUMMER BROOD INTERRUPTION FOR VITAL HONEY BEE COLONIES

... to watch the video for method application

Courtesy of the Nature Conservation Programme in North Macedonia – NCP[1]

QUEEN CAGING

- all brood and bees stay in the hive
- quick and easy
- opportunity for comb exchange
- opportunity for queen exchange
- long treatment period
- queen should be traceable
- special queen cage required

At a glance

For this brood interruption method, a special kind of queen cage is used to prevent the queen from egg-laying. Unlike the usual introduction or shipping cages, at least one sidewall of this type of cage consists of a queen excluder. This allows the worker bees free access to the queen and therefore ensures that the colony "feels" queenright. Simultaneously, the queen is stopped from egg-laying, producing an artificially induced broodless state and exposing the phoretic mites for effective treatment.

[1] "Alternative methods for varroa control in honey bee colonies", Uzunov and Büchler, 2020, Nature Conservation Programme in North Macedonia – NCP.

Basic idea

We know about the mite-suppressing effect of a brood interruption as part of natural swarming or as the natural end of brood rearing, but this usually isn't effective enough under field conditions when colonies suffer high mite burdens in late summer. So, it is quite convenient that oxalic acid, one of the most widely used organic acaricides, is only effective in broodless periods. By using this caging method, brood interruption and the oxalic acid application may go hand in hand and therefore result in a truly effective varroa treatment.

Various cages are available on the market, but it is also possible to build them yourself (Figure 17). The critical point to remember is that the bees need free access to the queen, and therefore the cage should be inserted in one comb and have excluder sidewalls. The cage should be inserted in an already drawn, empty comb by cutting a suitable hole between the comb wires. The cage should fit tightly inside the hole thus made, but if that is not the case, e.g., if you have accidentally cut out a slightly bigger hole, the bees will fix this in no time.

Figure 17. Alternative homemade queen cage (photo: Irakli Janashia).

You can prepare these "cage combs" at home and take them to the apiary, or you can just take a suitable comb from the hive and install the cage on site. Because you have to catch and cage the queen to apply this method, it is vital that you are able to find her. Remember that a marked queen will make your work a lot easier and see page 57 for further hints and tips for finding the queen.

SUMMER BROOD INTERRUPTION FOR VITAL HONEY BEE COLONIES

CREATING STRONG AND HEALTHY COLONIES WITH MINIMAL EFFORT!

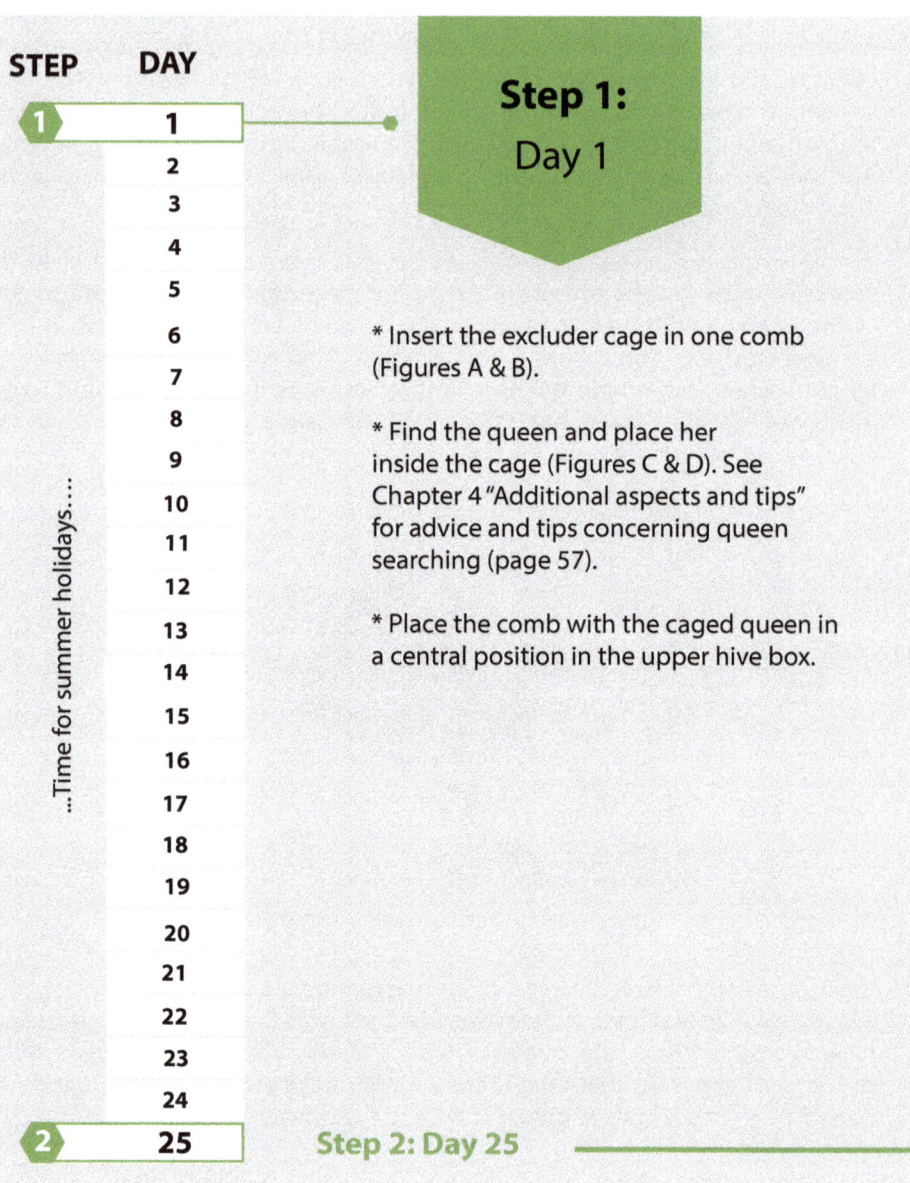

Step 1: Day 1

* Insert the excluder cage in one comb (Figures A & B).

* Find the queen and place her inside the cage (Figures C & D). See Chapter 4 "Additional aspects and tips" for advice and tips concerning queen searching (page 57).

* Place the comb with the caged queen in a central position in the upper hive box.

...Time for summer holidays...

Step 2: Day 25

Step 2: Day 25

* Exchange old and dark combs if necessary.

* Release the queen (Figure E) or exchange her with an new one in an introduction cage sealed with candy.

* Treat the colony with oxalic acid according to your local legislation (check the following pages).

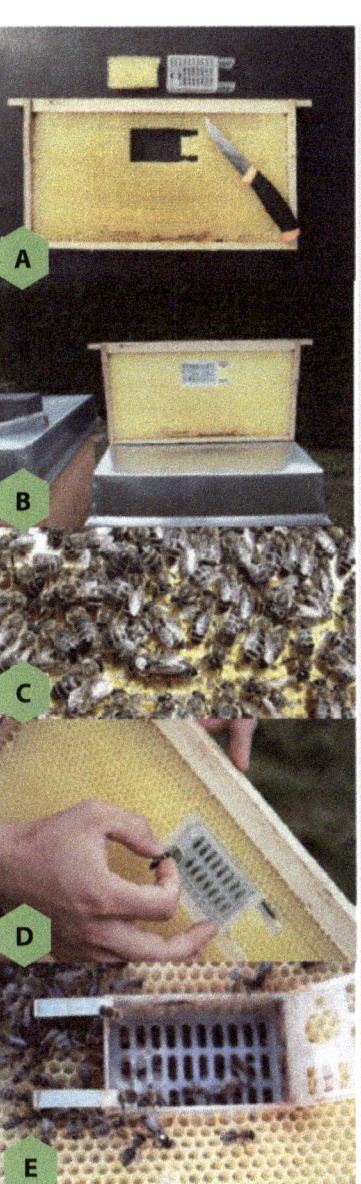

Oxalic acid – safety issues

Besides general work safety matters such as the avoidance of back injuries caused by too heavy loads or unsuitable working positions (e.g., insufficient height of the hive stand), the application of oxalic acid (technically oxalic acid dihydrate, in the following: oxalic acid) carries additional risks for the beekeeper.

Regardless of the type of application (trickling, spraying, or sublimation), any inhalation or contact of the agent with the skin, eyes, or clothes must be avoided at all cost. For self-protection, use gloves, glasses, and respirator masks suitable for the chosen application method (Figure 18). In addition, always take the wind direction into account (especially when you are spraying or sublimating) and make sure that you have the possibility of washing spilt acid away (e.g., with a bucket of water).

Figure 18. Self-protection equipment and tools needed for the application of oxalic acid solution. The yellow bucket contains water for washing away any spilt acid.

Application of oxalic acid

Oxalic acid is the most effective varroacide from the group of organic acids, which also includes lactic and formic acid. Its high efficiency is only gained in broodless periods and on phoretic mites, which makes this substance suitable for treatments in naturally broodless colonies (e.g., for critical or acute treatments during the winter) or in combination with the methods of brood interruption or brood removal described above.

Nevertheless, you should always follow the relevant laws and recommendations of your region or country. The permitted application types of the active compound as well as of distinct, licensed products vary between countries and are constantly being revised. This makes sense, since the right dosage and suitable application form depend on the respective area's environmental and biotic factors. Thus, please keep yourself updated on new legal revisions and scientific findings for your country. However, below we describe the most common forms of oxalic acid application and their respective pros and cons.

Basically, three application methods are suitable for oxalic acid treatments:

TRICKLING

In colder climates, the trickling of an aqueous solution of oxalic acid and sugar is a common method for critical treatments in the wintertime. Due to its low cost and time-saving application, this method is ideal for treating dense winter clusters where the active compound is well spread among the bees (Figure 19). However, application by trickling is also possible during the summertime, i.e., in broodless periods induced by brood interruption methods.

Figure 19. Trickling oxalic acid and sugar solution onto bees.

On a practical note, it is essential that you trickle the solution on the "streets" with bees (i.e. not on the top bars) and try to disperse it as far as possible if the cluster of bees is sitting over both supers. An application time when most of the bees are present in the hive (early morning or late evening) might improve efficacy.

SPRAYING

Under warmer temperatures, application by trickling might not always be the most efficient, because the bees are not sitting in a dense cluster but are distributed more widely over the combs. In order for the active compound to reach all bees – or rather, all mites – it is also positive to spray an aqueous solution of oxalic acid (Figure 20). The bees should be covered by a fine mist of droplets rather than be soaking wet (Figure 21)! Spray each comb with bees from below at a slightly diagonal angle to avoid spraying inside the cells. An advantage of this application method is that it is easier to adapt the dosage as the number of bees is checked comb by comb, which can be specifically important for the weaker colonies.

Figure 20. Spraying of oxalic acid solution after a brood interruption through caging the queen for 25 days.

Figure 21. Fine droplets of oxalic acid on bees.

SUBLIMATION

Another alternative for spreading the active compound throughout the colony is the sublimation of crystalline oxalic acid. In this case, the solid is heated by an electric heating plate or gas burner in or under the hive until it sublimates (Figures 22 & 23). The sublimate then wafts through the interior of the hive, where it "tackles" the mites. This method can be time consuming and dangerous for beekeepers, as they will have to heat the crystalline solid. A safer way of treating the colonies in warm conditions is the spraying technique, as mentioned above, but also time consuming. On the other hand, various studies have found the sublimation method to be quite effective and is accompanied by low bee mortality after treatment.

Figure 22. The hive body and bottom board should be sealed to minimise any leakage of the sublimated oxalic acid.

Figure 23. For the sublimation of crystalline oxalic acid (left) special equipment is needed (right).

TRAPPING COMB

... to watch the video for method application

Courtesy of the Nature Conservation Programme in North Macedonia – NCP[1]

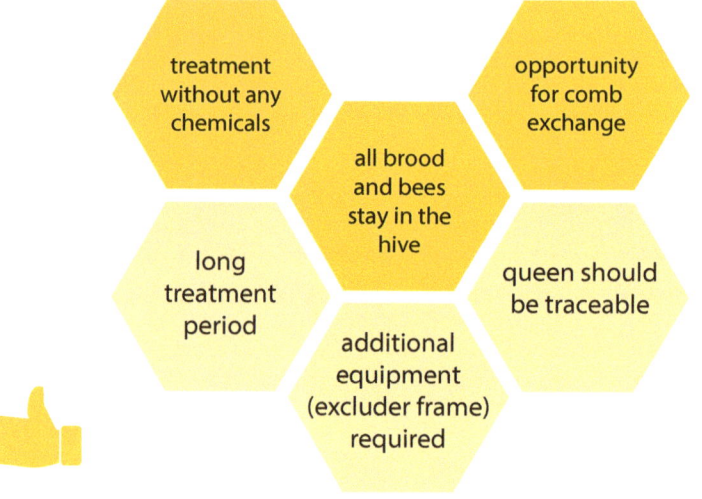

- treatment without any chemicals
- opportunity for comb exchange
- all brood and bees stay in the hive
- long treatment period
- queen should be traceable
- additional equipment (excluder frame) required

At a glance

To apply the trapping comb technique, a special excluder frame is needed. This kind of frame is available in beekeeping shops, but is also easy and fast to build. The minimum of tools and material required can be found in most beekeeping households. If you want to make your own excluder frame, please follow the step-by-step manual (see page 43), and you will be ready in a few minutes. As with the brood removal method, using this excluder frame the mites can be trapped in brood combs in order to remove them from the hives without using chemicals.

[1] "Alternative methods for varroa control in honey bee colonies", Uzunov and Büchler, 2020, Nature Conservation Programme in North Macedonia – NCP.

Figure 24. Homemade excluder frame for the trapping-comb technique. Similar frames are also available on the market.

Basic idea

The trapping comb technique is basically a combination of the brood removal and the queen caging approaches. Using an excluder frame (Figures 24 to 26) or a vertical queen excluder (Figures 27 & 28) in one box brood chambers like Dadant hives, the queen is confined on a single or limited number of combs to limit her egg-laying activity. Due to the schedule used in the application of this method, all adult mites looking for brood are forced into the trapping comb since all other brood cells will have hatched after the third week of treatment. After the capping of the brood cells in this trapping comb, it fulfils its purpose as a real "mite trap," because you can remove it with all of the mites trapped within.

Figure 25. Punching a hole in the trapping comb. This serves as a passage for the queen and allows her to switch between the two sides of the comb. Already dark brood combs are ideal, since the queens prefer them for egg-laying.

Figure 26. Left: Let the queen slip through the hole on the already closed side of the excluder frame. This ensures that you avoid harming her accidentally when closing the front side of the excluder frame (right photo). The timing of all the steps shown here is given in the step-by-step manual on pages 40 to 41 (Figure 29).

Figure 27. Vertical queen excluder in a single box brood chamber e.g., Dadant hive, here separating the two leftmost combs.

Figure 28. The entrance of the restricted section of the hive should be closed to prevent the queen moving to the other part.

FIGHTING THE MITE WITHOUT ANY CHEMICALS!

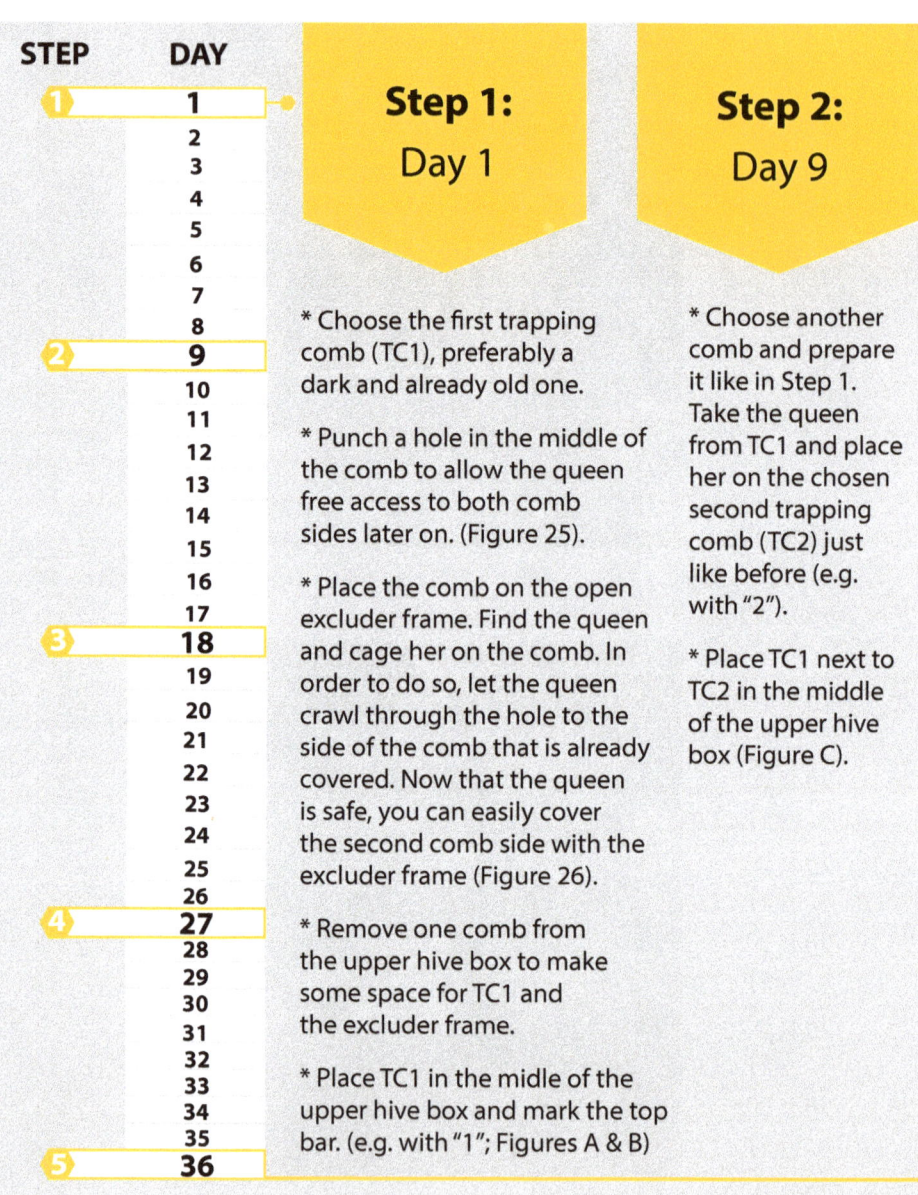

Step 1: Day 1

* Choose the first trapping comb (TC1), preferably a dark and already old one.

* Punch a hole in the middle of the comb to allow the queen free access to both comb sides later on. (Figure 25).

* Place the comb on the open excluder frame. Find the queen and cage her on the comb. In order to do so, let the queen crawl through the hole to the side of the comb that is already covered. Now that the queen is safe, you can easily cover the second comb side with the excluder frame (Figure 26).

* Remove one comb from the upper hive box to make some space for TC1 and the excluder frame.

* Place TC1 in the midle of the upper hive box and mark the top bar. (e.g. with "1"; Figures A & B)

Step 2: Day 9

* Choose another comb and prepare it like in Step 1. Take the queen from TC1 and place her on the chosen second trapping comb (TC2) just like before (e.g. with "2").

* Place TC1 next to TC2 in the middle of the upper hive box (Figure C).

Step 3: Day 18

* Choose another comb and prepare it like in Step 1. Take the queen from TC2 and place her on the chosen third trapping comb (TC3) and mark the top bar (e.g. with "3").

* Place TC2 next to TC3 in the middle of the upper hive box (Figure D).

* Exchange the now capped TC1 it with a new comb or foundation. Melt the trapping comb TC 1 to harvest the wax.

Step 4: Day 27

* Exchange all old and dark combs (including the now capped TC2) with new combs or foundation.

* Remove the excluder frame and release the queen.

* Mark the top bar of TC3 (e.g. with "3").

* Leave TC3 in its former position in the upper hive box (Figure E).

* Remove and melt the now capped TC3. Exchange it with a new comb or foundation (Figure E).

* Take care of the colony's food stocks, especially if you used foundation.

Step 5: Day 36

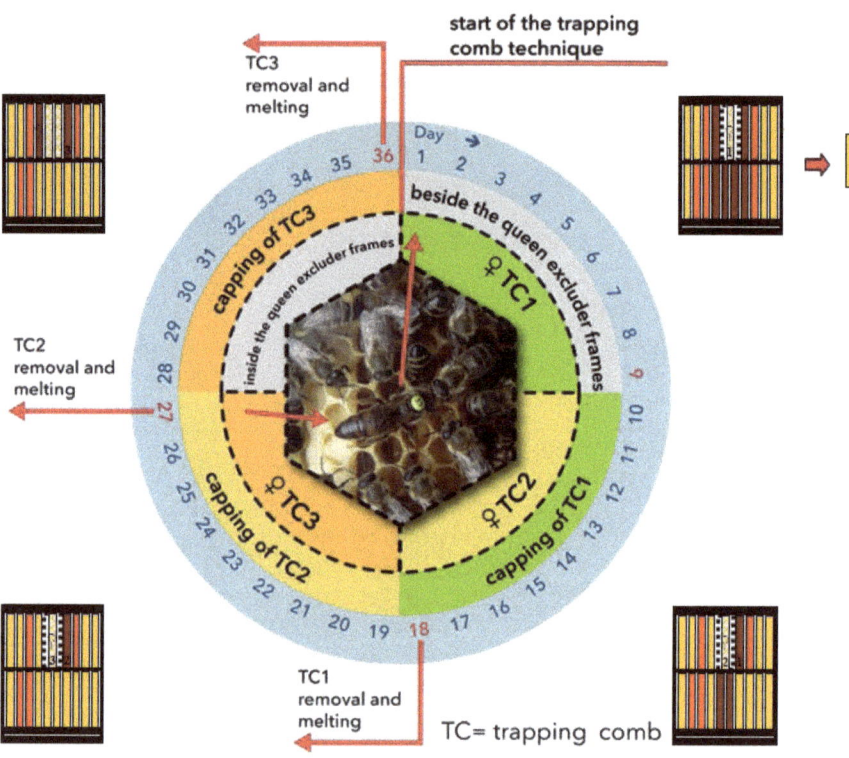

Figure 29. Time wheel chart hints for the trapping-comb technique. The outer circle shows the day of treatment, the middle circle indicates the stage of the trapping combs, while the inner circle indicates on which trapping comb the queen is for each date (modified after LLH, Bee Institute in Kirchhain, Germany, 2018). Instead of 3 x 9-day intervals, you may also work in 4 x 7-day intervals to cover a period of 28 days.

Making a trapping comb

The minimum of tools and materials which are required can be found in most beekeeping households. Just take a look at Figures 30 to 36, and you will be ready in a few minutes.

Required equipment:
- Wooden cleats or strips of approximately 10 x 10 mm (length depends on the size of your frames)
- Plastic excluder material cut to size also depending on frame size.
- Approx. 20 cm of frame wire.
- Four nails.
- Approx. 10 cm of plastic webbing.
- Side-cutting pliers or other cutting tool.
- Stapler and staples (or hammer and nails).

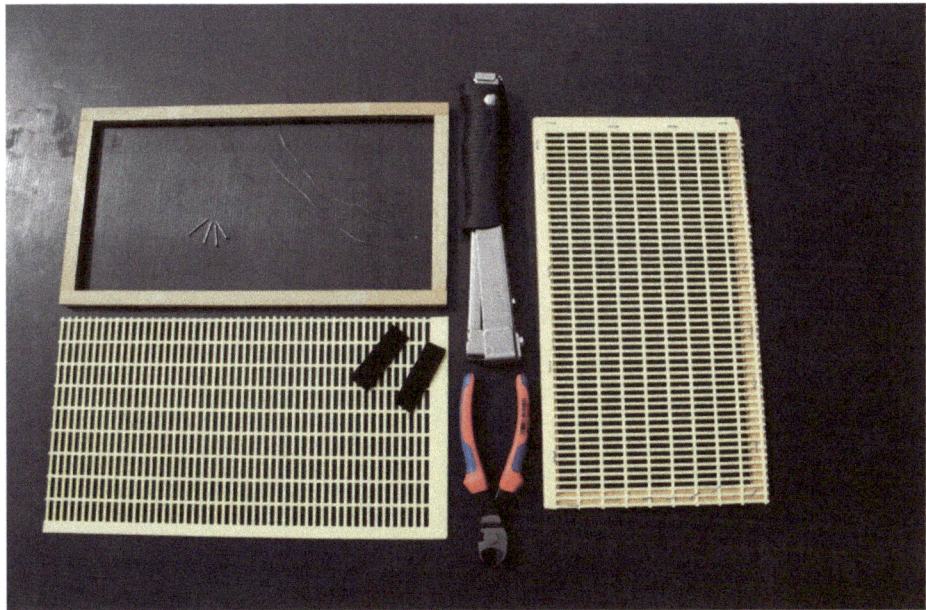

Figure 30. Tools and materials needed to build an excluder frame.

Step 1: Cut the cleats according to the size of your frames. Build two identical frames covering the entire comb surface as well as the top and bottom bars and fitting between your spacers. Be careful not to split the wood putting in the nails since the cleats are thin. To avoid having to nail in the end grain, you may also choose a more stable connection such as half-lap joints.

Step 2: Cut the excluder material into two pieces according to the outer dimensions of your newly built frames and attach one piece to each of the two frames using staples or nails (Figures 31 and 32).

Step 3: Cut the webbing into two pieces and fix them with staples or nails onto one excluder frame. Hold both excluder frames against a comb to measure the correct length of the webbing. According to this measurement, attach the loose ends of both webbing strips to the other excluder frame (Figure 33).

Step 4: Place a nail in each of the left and right edges of the upper cleats of both excluder frames so that two nails lie opposite when the excluder frames are attached to a comb (Figure 34). Cut two pieces of frame wire (approx. 10 cm each) and wrap one end of each piece around a nail on the left and right side of the excluder frame (Figures 35 and 36).

Step 5: The frame is now complete and ready for use.

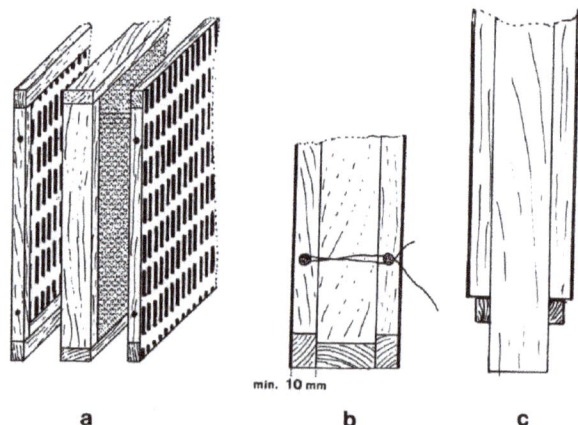

Figure 31. a) comb with two excluder frames; b) simple fastening of the single frames with threads of comb wire; c) view from the top: attached excluder frames between the sidebars of "Hoffman" frames. "Hoffman" frames have broadened sidebars as spacers (modified after Maul *et al.*, 1988).

Figure 32. Stapling the excluders to the cleat frame.

Figure 33. Connecting both frames with webbing.

SUMMER BROOD INTERRUPTION FOR VITAL HONEY BEE COLONIES

Figure 34. Putting a nail in each of the upper corners.

Figure 35. Wrapping the comb wire to two nails of one frame side.

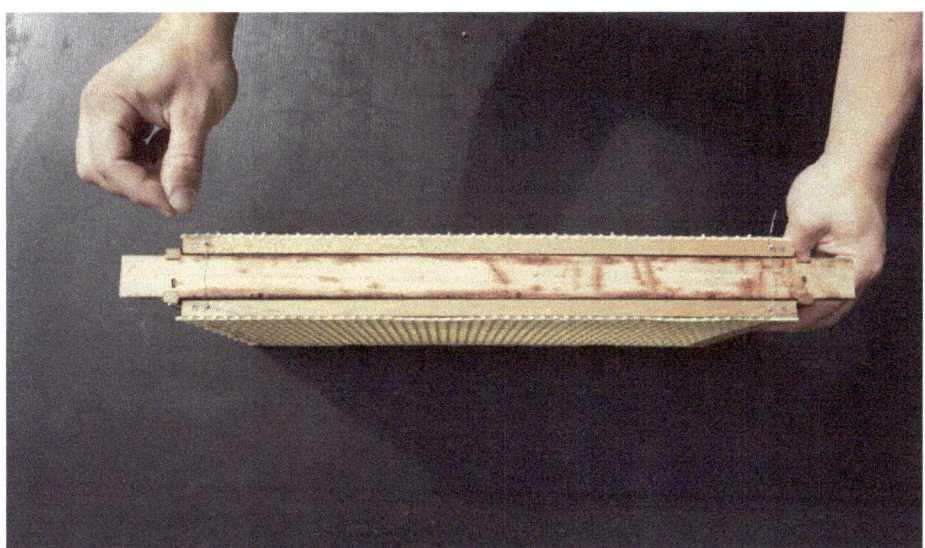

Figure 36. Attaching the excluder frame to a comb to check that it fits.

There are many alternatives designed by beekeepers and some of them are available on the market (Figure 37).

Figure 37. An excluder frame made from metal.

3 FACTORS AFFECTING APPLICABILITY

FACTORS AFFECTING APPLICABILITY OF THE METHODS

Experience

The beekeeper's experience is a critical factor in choosing the right method. By "experience," we mean knowledge about honey bee and mite biology and practical experience with beekeeping techniques as well as the knowledge of ecological factors such as climate and food sources for honey bees. Thus, an experienced beekeeper will easily be able to foresee most of the challenges and the effort and resources necessary for the implementation of each of these brood interruption methods.

On the other hand, a newcomer to beekeeping may also straightforwardly accept the approach, because they may not be primarily concerned about their beekeeping economy and may also wish to strive towards a "near-natural" production of honey bee products. Nevertheless, the best efficiency and advantage of using these methods can be expected from beekeepers with at least a few years of beekeeping experience.

The region, season and environment

As noted above, knowledge and experience about the honey bee's biology and its environment are crucial for obtaining the best results from these methods. Not all regions or environments are equally suitable for practising brood interruption as a varroa control measure. Regions with a long brood period are more suitable, because the bees will have sufficient time to "recover" from the brood interruption in their colonies before the onset of winter. At the same time, the relevance of an artificial brood interruptions increases with the duration of continuous brood development in order to limit the exponential growth of the mite population. The results from some published studies and our own experience show that as well as southern European regions, the methods can be successfully used in northern Europe, above 50° of latitude (see "Further reading"; page 77).

Since variation between years may be significant, using a calendar will not help to decide when to apply any of these techniques. Instead, good indicators for finding the most suitable period for their use will be the swarming season and the time of the expected final honey harvest. Thus, beekeepers should use these indicators as a guide to decide when the methods can be best applied. In general, as described above, the time of the end of the last summer honey flow is usually the most suitable treatment period. Following these guidelines, the colonies will have enough time to rebuild to sufficient strength for proper overwintering and can still be used for honey harvest in advance.

Size and type of beekeeping operation

The high efficiency of the brood interruption methods in ensuring a healthy winter bee development and its potentially positive effects on colony strength and honey productivity, makes it attractive for all kind of beekeeping operations. Nevertheless, small scale and hobby beekeepers may be more flexible with regard to the demanded work load and timing. In addition, their beekeeping pretty much depends upon the availability of their free time when weekends and afternoons are suitable for work on the bees, for instance, 4 x 7-day intervals compared to 3 x 9 for the trapping comb. This allows for a lot of flexibility regarding their integration into a regular work / life pattern. However, this may not be the case for commercial beekeeping operations, where the time needed, the effort, and the complexity of these procedures may significantly affect the overall beekeeper's revenue. Though, queen caging with the use of oxalic acid is a good example of how such a method can be integrated even in commercial beekeeping operations. In Italy and Germany, there are beekeepers practising this or a slightly modified method in their commercial-scale beekeeping operations (Figure 38).

Figure 38. Queen caging as varroa control method is practiced in many commercial beekeeping operations. Upper: in Italy (apiary of commercial beekeeper Marco Moretti; photo Marco Pietropaoli) and Lower: in Germany (LLH, Bee Institute Kirchhain).

Figure 39. Performance testing apiary at the Bee Institute in Kirchhain, Germany.

The same techniques can even be successfully integrated into populations undergoing performance testing, for example, in breeding programmes aiming to produce varroa resistant stock (Figure 39). The total brood removal method can be used in selection schemes for breeding varroa resistant bees, particularly for drone-producing colonies at the mating stations. By applying the method, the breeders manage to keep mite infestation in drone colonies at subcritical levels without the use of medication, thus enabling the drones from the fittest colonies to mate with the virgin queens. In this way, natural selection elements, such as high mite infestation and pressure on the drones' health, can be incorporated into the selection schemes (Figures 39 & 40).

Figure 40. Drone-producing colonies at the resistance mating station Gehlberg in Germany.

Time requirements, workload and risks

Each of the brood interruption methods has a different requirement for time and physical effort for preparation and application. As with any advanced practice, there are many pros and cons. No one method is clearly better than another, and the beekeeper has to choose what fits best for their operation's size, working schedule, work habits etc. However, based on our practical experience, we have ranked them in the summary table (Table 1), based upon time requirements and workload.

SUMMER BROOD INTERRUPTION FOR VITAL HONEY BEE COLONIES

Method	Use of chemicals	Optimal timing*	Time requirements**	Workload	Experience required	Additional person needed	Type of beekeeping operation	Comment
Brood removal	no	0 to 12	medium (25–35 min. per colony)	medium-high	medium-high	Recommended but not obligatory	mostly hobby, commercial in case of hive multiplication	Risk of starving/robbing, need for feeding
Queen caging	yes	15 to 25	low (15–20 min. per colony)	low	low-medium	no	all types	Queen must be caught (possible loss)
Trapping comb	no	15 to 35	high (30–40 min. per colony)	medium-high	medium-high	no	hobby	Queen must be caught (possible loss)

*days before final harvest   **including all steps and manipulations.

Table 1. Summarised comparison of the methods with pros and cons ranked by time requirements, workload and other parameters.

Queen caging is the most straightforward and least laborious method compared to the trapping comb and total brood removal methods. This is why, aside from the small-scale beekeepers, the method could also be attractive for commercial operations. Its main drawback is the use of a chemical substance (oxalic acid) to kill the mites, which, if misapplied, can also be hazardous for the bees and the beekeeper. Beekeepers' main concerns, however, tend to be the effects of caging on the queens. However, our experience and the results from our studies show that during the caging as well as afterwards, queen survival is not affected, and any losses are within the limits of the biologically expected loss rate. The effects of caging on the queen's laying performance could be further studied, even if a reduction and interruption of egg-laying activity in the course of swarming is part of the colony's natural life cycle.

The use of a trapping comb is a well-designed approach for removing mites from the colony without the use of any chemical. On the other hand, the method requires special equipment (the comb cage) and more visits to the apiary than the remaining two methods. Since there is no use of chemicals, it is not harmful to human or bee health. It seems that queens do not suffer any adverse effects on their survivability, life expectancy, or performance during captivity and after their release.

The brood removal technique is the most laborious method and requires spare equipment such as frames, combs and foundations, possibly bee food, hive boxes, etc. Help from another experienced person is also useful for the efficient application of the method. The main challenge associated with this method is the use and management of the frames removed, which sometimes can be numerous. However, it is a non-chemical method, with comparatively low risk of queen loss and several additional benefits. If well planned and synchronised with queen production, the method can not only be used for replacing queens and combs, like the treatments described above, but can also be used for colony multiplication (i.e., the production of additional nucs late in the season; Figure 41).

Figure 41. Brood collectors overwintered and developed into full-production colonies in the following season.

4 ADDITIONAL ASPECTS AND TIPS

ADDITIONAL ASPECTS AND TIPS

FINDING THE QUEEN

In two of the brood interruption methods, queen caging and the trapping comb, the queen must be found and placed in the cage or trapping frame to apply the method. On the one hand, this could be considered a disadvantage of these methods, as it requires certain skills and experience from the beekeeper. On the other hand, finding the queen is an essential and handy skill in many beekeeping techniques, so it is worth investing a little time to learn these skills to save much time later. In addition, having the queen "in hand" or confined in the cage provides a good possibility for exchanging / replacing her after the methods' application. Although it can sometimes be tricky to find the queen, there are some useful tips for finding the queen quickly and easily, even in a large colony:

Marking the queen

The best way to save a lot of your valuable time as a beekeeper is to invest a little time to mark the queen. That can be done during the rearing process but also during any hive inspection when the queen is observed. In small mating colonies (mating boxes or nucs), it is especially easy to find the mated queen. Before you introduce her to the new colony, mark her with a numbered opalite disc or a colour pen (Figure 42). This is an essential practice, since at the time of methods application the production colonies are strong and populous. Thus, do not introduce any unmarked queen into your hives. Marked queens, regardless of whether self-raised or purchased from another beekeeper, are the key point for easy identification of the queen in the colony and, therefore, for many beekeeping techniques (Figure 43).

  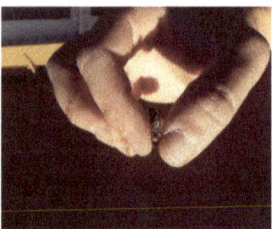

Figure 42. Find and fix the queen on the comb (left), carefully catch her (centre) and hold and mark with the corresponding colour for the year of birth (right).

Figure 43. Marked queen on a comb with many recently laid eggs.

Check the brood area

Apart from the mark on the queen's back, you can also use other "signs" to find her. Like any good tracker, you may find her based on the traces she has left. Of course, her footprints will be too hard to find, but what about eggs and young larvae? You will find the queen in most of the cases on combs containing young brood (Figure 43).

There is often too much brood in the colony to restrict the search area based on the position of the brood nest during peak brood season. If you can't find the queen because brood is present nearly everywhere, you may use this trick:

Place a queen excluder between the lower and the upper brood chamber (Figure 44). After three days, you will find eggs only in the box which harbours the queen. In this way, you can reduce your search area by up to 50 % and find the queen much more easily.

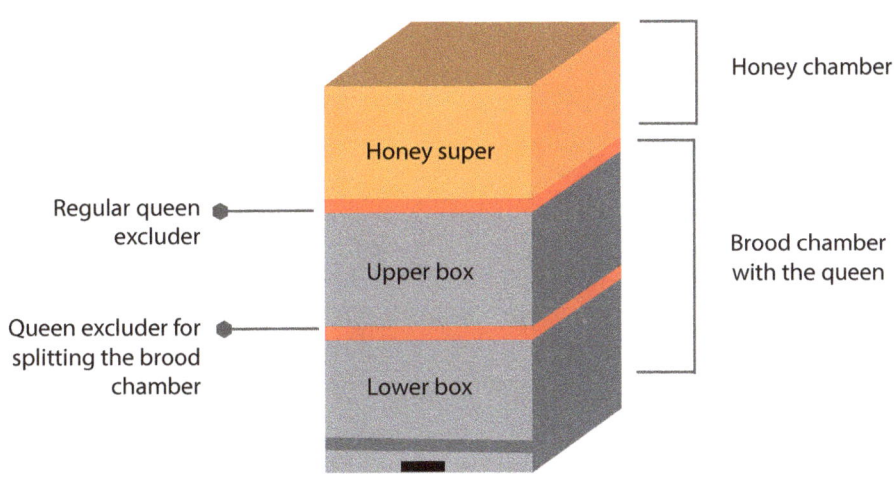

Figure 44. Placing a queen excluder between the brood boxes for three days will help the beekeeper to reduce their efforts for finding the queen.

Search for her systematically

Sometimes beekeepers trust their gut feeling to find the right comb when searching for the queen. In most cases, this is very time consuming and inefficient. First of all, you should split the boxes to prevent the queen from changing her position while searching for her (Figure 45) and only carefully use smoke in order not to startle her.

Figure 45. Split the boxes and search for the queen in each separately.

Take care! Sometimes the queen is also sitting on the queen excluder or the lid (Figure 46). Remove the outer comb (most probably with food) and place it beside the hive. In exceptional cases, she could also be sitting on this comb or the hive wall. Make sure to create a gap between the combs, which makes your work a lot easier and prevents the queen from "escaping" to combs you have already checked.

Figure 46. Careful handling of frames and hive parts as well as suitable bee space in the box should prevent such unfortunate accidents! In this case, the space between top-bars and lid was too narrow.

Begin searching for her on the next comb (probably already containing brood cells) and work your way through the hive box by checking the combs one by one. After checking one comb, place it on the "already checked" side of the hive (Figure 47). Like this, there is always a gap between the combs you already checked and the combs you haven't checked so far.

Since queens avoid light and try to escape to darker areas, it makes sense to already have a look at the visible side of the next comb while taking out the previous one for closer inspection. If the queen is sitting on this comb side, you will see her before she is able to escape to the inner side of the comb.

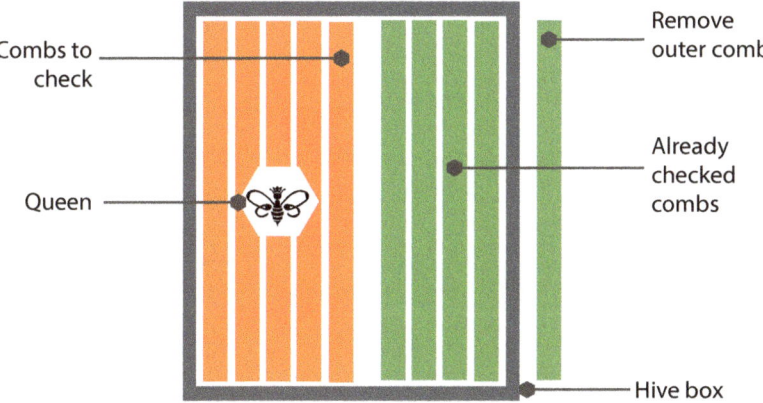

Figure 47. Check the combs one by one carefully for the queen, and always keep a gap between the already checked combs and those that still need to be checked.

If you don't find her in the upper box, continue in the same manner with the lower one. If you have checked each box twice and still haven't found the queen, the colony will usually get too nervous for proper searching. In this case, it is better to apply the separation trick with the queen excluder described earlier (see "Check the brood area"; page 58).

COMB REPLACEMENT – FROM DARK TO LIGHT

As is well known to beekeepers, older and already incubated combs get darker with each brood cycle. The colour changes due to residues (silken fibres and exuviae) and faeces of the larvae, also the size of the cells becomes smaller with each layer of residues, resulting in smaller bees hatching from them. Such "dirty" combs are breeding grounds for pathogens and should be removed from the hives (Figure 48). Many beekeepers struggle with this task, because old brood combs commonly contain vast amounts of brood which they avoid removing in the summertime. On the other hand, the colonies lack both workforce and motivation to build up new combs from foundation in broodless periods (very early or late in the season). In some regions brood is present mostly all year round, making it even more difficult to renew the combs without sacrificing brood cells. Here the brood interruption methods for varroa control score another point!

Each method enables the beekeeper to replace old combs and, therefore, to create a healthy and hygienic nesting site for his bees while harvesting valuable wax. After all brood is hatched during application of the trapping comb or caging methods, the beekeeper can easily replace the old combs with new ones built from honey supers or from foundation.

Figure 48. Such old combs serve perfectly well for trapping combs, but should be removed from the hive at any cost.

Because the queen caging and trapping comb methods both include a moment when the colonies don't have any brood to care for by the many worker bees, they willingly build new combs out of foundation (Figure 49). Since wax production is a costly business for the bees, it is crucial to keep an eye on the nectar flow and to feed the colonies if necessary. The same applies to the total brood removal technique, but the replacement of old comb material here is already included. For the use of removed brood combs in brood collectors, see the description of the "Total brood removal" method (page 15).

Figure 49. Bright brood comb approximately one month after a complete brood removal. The colony has built up fresh combs from foundation, creating a nice new brood nest.

REQUEENING DURING APPLICATION OF THE METHODS

In some old beekeepers' jokes, techniques that are risky for the queen are described as "coming with a built-in requeening option", but that's not true for the methods described here! In addition to many other positive aspects, all of the methods described here offer good opportunities for requeening the colonies (Figure 50). Since a strong colony with brood doesn't easily accept a foreign queen in the high-breeding period, beekeepers usually requeen their colonies at the end of the season. Nevertheless, in case of a brood interruption with a period without or with limited areas of open brood, the colonies accept a newly introduced queen very well. Keep in mind that this phenomenon is also well known from shook swarms, where an artificial swarm of 1.5 – 2.5 kg of bees gets equipped with a new queen. By requeening your colonies like this, you don't have to keep mated queens in mating boxes until the end of the season and might use the boxes for another batch of queens.

To requeen the colonies, remove the old queen and place a young queen in an introduction cage sealed with candy in the hive's central position.

Figure 50. A young queen is one of the preconditions for a strong colony (photo: Marin Kovačić).

The best time for doing so depends on the method you choose:

Brood removal: you can change the queens immediately when removing the brood combs. Remove the old queen and replace her with a new one in an introduction cage (remember that the colony without brood combs is basically a shook swarm).

Queen caging: remove the old queen on day 25 instead of releasing her from the cage. Place a young queen in an introduction cage in the colony.

Trapping comb technique: just as in the queen caging method, you can replace the queen instead of releasing the old one at the end of the application period. In this case, as there may be open brood on TC 3, watch out for any emergency cells!

5

**FAQ &
PRACTICAL
EXPERIENCE**

FREQUENTLY ASKED QUESTIONS & PRACTICAL EXPERIENCE

What about queen losses and other incidents? Isn't it a huge risk?

This is a legitimate question that immediately comes to the beekeeper's mind: Isn't it hazardous for the queen and the whole colony to manipulate the brood nest during this time of the year? The answer is: No, there is nothing to worry about if you follow the recommendations and instructions for proper queen and colony handling. From our recent studies in Central and SE Europe (Gabel, 2016; Büchler and Uzunov, 2016; Uzunov and Büchler, 2020; Büchler *et al.*, 2020) it is evident that the colonies where the methods for brood interruption were applied did not experience significantly higher losses or supersedure of queens compared to the conventionally treated colonies. Contrary to expectations, the lowest queen losses occurred in the colonies where queen caging was applied. Similar results were obtained from other studies in Europe (Rivera-Gomis *et al.*, 2017), and it is worth mentioning that many large-scale beekeepers in Italy have been using the queen caging method as a regular (and sometimes as their only) strategy for varroa control. Moreover, if proper beekeeping practice is applied, beekeepers may further utilise the methods for replacing old and non-productive queens or making surplus nucs in case of brood removal (Figure 51).

Figure 51. Additional nucs gained from removed brood combs. Sometimes establishing of single box collectors fit better to beekeepers' management concept.

What about the colonies? Will they develop until the next winter?

Indeed, as always in beekeeping, timing is the critical factor. But one should keep in mind that on-time and properly applied methods will not negatively affect colony development for the next winter and spring seasons. In fact, quite the opposite is true, as colonies treated this way will rapidly "compensate" for the interruption of their brood rearing (Figure 52) and regularly approach the risky winter period in an even more robust condition than the colonies treated with chemicals to control varroa. Moreover, we found that the colonies with brood interruption overwintered successfully, similar to the colonies with conventional treatments (Gabel *et al.*, 2016). Subsequently, the colonies meet beekeepers' criteria for the spring period and reach their full power to exploit the coming nectar flow.

Why is that? Basically, this phenomenon is based on two factors. An increased brood activity as well as a general improvement of the colony's health. You will understand the former by imagining a swarm settled on foundation. Just like after a brood interruption or brood removal, the bees will rapidly build up a substantial but compact brood nest to compensate for the loss of brood cells.

Figure 52. Compact brood nest on fresh combs after brood removal, bees were fed with syrup in order for them to build up foundation.

The latter effect is connected to this "reboot" of the brood nest on new combs. During the method's application, the beekeeper may replace the old and deformed combs, which undoubtedly contribute to the colony's' health status, development and performance. In our studies (Büchler, 2007; Gabel, 2016), we found that brood interruption methods, beside the evident effect in varroa control, improve the colonies' overall health status by lowering loads of other pathogens such as viruses and *Nosema* spp. (Figure 53).

As becomes evident, the brood interruption techniques are a real all-round method for late-season beekeeping tasks: effective varroa control, the easy replacement of old combs and safe requeening. Even the honey yield from late nectar flows might be increased if the time of brood interruption and thereby a lower self-consumption of the colony is induced properly. Thus, even with a sudden interruption of the brood-rearing, "the bees will find a way".

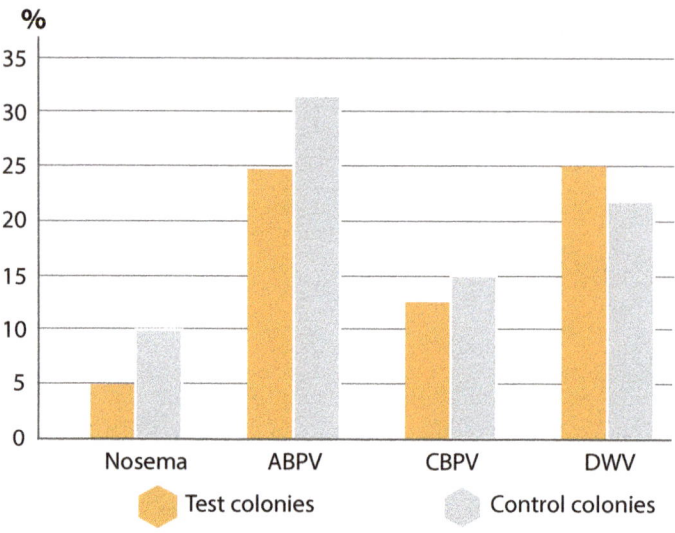

Figure 53. Infection frequency of bee samples of both comparison groups with the pathogens of *Nosema* spp., acute bee paralysis virus (ABPV), chronic bee paralysis virus (CBPV) and deformed wing virus (DWV), modified after Büchler, 2007.

How effective are the methods against varroa? Are they sufficient to keep my colonies safe?

Brood interruption methods successfully keep varroa infestation below critical levels, which in most parts of Europe are around 5 mites per 10 g (or approximately 100) bees. Our recent studies in Germany and Macedonia (Figure 54) show that the colonies treated with some of the brood interruption methods successfully controlled varroa infestation below 5%, even when those methods only are used year-round. This indicates that with adjustment to local environmental conditions, the methods work well in different parts of Europe. Still, a general recommendation is to keep an eye on the mite infestation by applying some of the varroa monitoring methods and if the critical levels are exceeeded, to treat the colonies.

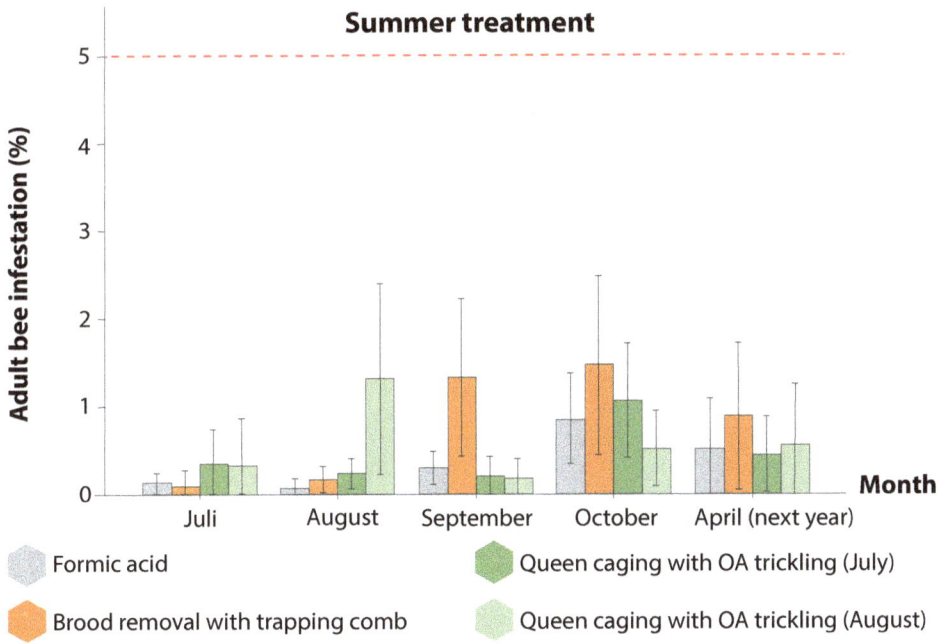

Figure 54. Bee infestation [%] of differently treated colonies after treatment, the dotted red line marks the general threshold of 5 % bee infestation (Gabel *et al.*, 2016). OA stands for oxalic acid.

In the recent Europe-wide study on the use of the methods for summer brood interruption (Büchler *et al.*, 2020), conducted in 15 apiaries in 10 European countries, we show that the method of queen caging, with the administration of 4.2% oxalic acid by trickling, was efficient by eliminating almost 90% of the varroa mites in the colonies. Comparable efficacy was also obtained by applying chemical-free brood interruption methods such as the trapping comb and brood removal methods.

But aren't those methods difficult to apply?

Application of the methods for brood interruption can be challenging and significantly depends on the beekeeper's experience, as well as on the scale of their beekeeping and the environmental conditions. The best advice we can give to the beekeepers is to test the methods on a small number of colonies, and once this experience is gained, the one that works best can be considered as a prime method for varroa control.

6 CHECKLIST

Brood removal

If you would like to use oxalic acid:

Equipment required for one colony		check	If you would like to use a trapping frame:
Oxalic acid solution	Depends on application method and colony strength		No additional equipment needed ✓
Sprayer or syringe	In general: one		
Gloves, goggles and mask	In general: one set per person		

If you would like to use the brood combs for nuc-building:

Equipment required for one colony		check ✓	If you prefer to melt the brood combs:
Complete empty hive	Approx. one per two colonies		No additional equipment needed ✓
Formic acid	Depends on application method and hive size		
New queen (virgin or mated queen)	One per nuc		

SUMMER BROOD INTERRUPTION FOR VITAL HONEY BEE COLONIES

Queen caging		
Equipment required for one colony		**check** ✓
Queen cage	One	
Frames with foundation or built combs	For replacement of old and dark combs (approx. 10)	
Beekeeping equipment (smoker, brush and hive tool)	In general: one set per person	
Right timing on two dates	Check your calendar!	

Trapping comb		
Equipment required for one colony		**check** ✓
Queen excluder frame	One	
Frames with foundation or built combs	For replacement of old and dark combs (approx. 10)	
Beekeeping equipment (smoker, brush and hive tool)	In general: one set per person	
Right timing on four dates	Check your calendar carefully!	

7 FURTHER READING

FURTHER READING

Büchler R. (2009) *Vital colonies thanks to complete brood removal*, ADIZ/db/IF 7/2009, 10-12. https://freethebees.ch/wp-content/uploads/2018/01/brood-removal-web.pdf

Büchler R., Uzunov A. (2016) *Mach mal Pause: Varroa-Bekämpfungsstrategie neu ausrichten!*, Imkerfreund, 03.2016, Deutschland.

Büchler R., Uzunov A. (2016) Selecting for varroa resistance in German honey bees. *Bee World*, 93(2): 49-52. https://doi.org/10.1080/0005772X.2016.1252178

Büchler R., Uzunov A., Nanetti A., Prešern J., Kovačić M., Charistos L., Coffey M F., Formato G., Gerula D., Rivera Gomes J., Malagnini V., Galarza E., Hatjina F., Vojt D., Nedić N., Panasiuk B., Pavlov B., Puškadija Z., Smodiš Škerl M., Vallon J., Wegrzynowicz P., Wilde J. (2019) Seasonal brood interruption as an effective measure for Varroa control. *46th Apimondia International Apicultural Congress Montréal, 8-12 September 2019*.

Büchler R., Uzunov A., Kovačić A., Prešern J., Pietropaoli M., Hatjina F., Pavlov B., Charistos L., Formato G., Galarza E., Gerula D., Gregorc A., Malagnini V., Meixner M.D., Nedić N., Puškadija Z., Rivera-Gomis J., Rogelj Jenko M., Smodiš Škerl M.I., Vallon J., Vojt D., Wilde J., & Nanetti A. (2020) Summer brood interruption as integrated management strategy for effective varroa control in Europe. *Journal of Apicultural Research*, 59(5): 764-773. https://doi.org/10.1080/0 0218839.2020.1793278

Gabel M., Uzunov A., Büchler R. (2016) *Alternative Sommerbehandlung der Honigbiene (Apis mellifera) gegen* Varroa destructor *(Anderson & Trueman)*, 63. Jahrestagung der AG der Institute für Bienenforschung e.V., 22-24 March 2016, Braunschweig, Germany.

Gabel M., Uzunov A., Büchler R. (2016) Summer brood interruption for vital honey bee colonies (results and experience from a study in Germany). *COLOSS Varroa Control Task Force Workshop, 19-20 May 2016, Unije, Croatia*.

Gabel M., Büchler R., Uzunov A. (2016) *Mach mal Pause – Versuchsergebnisse zur Brutunterbrechung stimmen optimistisch*, ADIZ • die biene • Imkerfreund, 9-11, 07.2016.

Gabel M., Uzunov A., Büchler R. (2016) Brood interruption and brood removal as alternative summer treatments of honey bees (*Apis mellifera* L.) against *Varroa destructor* (Anderson & Trueman), GFÖTagung, 5-9 September, Marburg, Germany.

Gabel M., Büchler R., Uzunov A. (2017) Biotechnical approaches for varroa control – different applications of brood interruption, brood removal and trapping combs in colony management. *45th International Apicultural Congress – Apimondia 2017, 29 September – 04 October 2017, Istanbul, Turkey*.

Gabel M., Uzunov A., Dreher C., Büchler R. (2017) *Quodlibet – guidance to practical and effective biotechnical Varroa control* (Quodlibet – Leitfaden zu praktikabler und effektiver biotechnischer), AG-Tagung 2017 in Celle, Germany.

Gabel M., Uzunov A., Wallner K., Büchler R. (2017) „Mach mal Pause", Untersuchungen zu Brutpause und Bannwabe gehen in die nächste Runde. *Bienen & Natur*, 14-16, Issue 7, 2017, Germany.

Goodwin M., Van Eaton C. (1999) *Control of varroa - A guide for New Zealand beekeepers*. New Zealand Ministry of Agriculture and Forestry. ISBN: 9780478079586

Kovačić M., Uzunov A., Tlak Gajger I., Pietropaoli M., Soroker V., Adjlane N., Benko V., Charistos L., Dall'Olio R., Formato G., Hatjina F., Malagnini V., Freda F., Otmi A., Puškadija Z., Villar C., Büchler R (2023) Honey vs. mite - A trade-off strategy by applying summer brood interruption for *Varroa destructor* control in the Mediterranean Region. *Insects*, 14: 741. https://doi.org/10.3390/insects14090751

Matthes, D. (1978). Tiersymbiosen und ähnliche Formen der Vergesellschaftung. Fischer, Stuttgart, New York, ISBN: 978-3437201936

Maul V., Klepsch A., Assmannwerthmuller U. (1988) The trapping comb technique as part of bee management under strong infestation by *Varroa jacobsoni* Oud. *Apidologie*, 19, 139-154. https://doi.org/10.1051/apido:19880204

Meglic M., Augustin V. (2007) *Varoja, Cebela, Cebelar*. Cebelarska zveza Slovenije. Lukovica. ISBN: 9616516094 / 9789616516099

Meixner M., Uzunov A., Büchler R. (2017) *Estimating regional varroa threshold levels across Europe*. COLOSS Virus Task Force Workshop, 6-7 April 2017, Avignon, France.

National Bee Unit (2017) *Managing varroa*. The Animal and Plant Health Agency, UK.

Nanetti, A., Büchler, R., Uzunov, A., Gregorc, A. (2016) *Coloss - Varroa Control Task Force brood interruption study 2016/2017*. http://www.coloss.org/taskforces/varroacontrol/protocols-brood-interruption-varroa-task-coloss-2016-jan-final

Nazzi F., Le Conte Y. (2016) Ecology of *Varroa destructor*, the major ectoparasite of the western honey bee, *Apis mellifera*. *Annual Review of Entomology*, 61: 417-432. https://doi.org/10.1146/annurev-ento-010715-023731

Pietropaoli, M., Giacomelli, A., Milito, M., Gobbi, C., Scholl, F., Formato, G. (2012) Integrated pest management strategies against *Varroa destructor*, the use of oxalic acid combined with innovative cages to obtain the absence of brood. *Eur. J. Integr. Med.*, 15: 93. https://doi.org/10.1016/j.eujim.2012.07.691

Ramsey, S. D., Ochoa, R., Bauchan, G., Gulbronson, C., Mowery, J. D., Cohen, A., Lim, D., Joklik, J., Cicero, J., Ellis, J., Hawthorne, D. & van Engelsdorp, D. (2019). Varroa destructor feeds primarily on honey bee fat body tissue and not hemolymph. Proceedings of the National Academy of Sciences, 116(5), 1792-1801. https://doi.org/10.1073/pnas.1818371116

Rosenkranz, P., Aumeier, P., Ziegelmann, B. (2010) Biology and control of *Varroa destructor*. *J. Invertebr. Pathol.*, 103: S96–S119. https://doi.org/10.1016/j.681jip.2009.07.016

White, P. S., Morran, L., & de Roode, J. (2017). Phoresy. *Current Biology*, 27(12): R578-R580. https://doi.org/10.1016/j.cub.2017.03.073

Uzunov A., Büchler R. (2014) *Potential of brood removal method for sustainable varroa control*. 21-22 May 2014, Bled, Slovenija.

Uzunov A., Büchler R. (2014) *Brood removal method - positive effects on honey bee colony health and performance.* COLOSS Workshop "Varroa control strategies", 2-3 December, Bologna, Italy.

Uzunov A., Büchler R. (2016) *Zwangsbrutpause zur Varroabehandlung.* Jahresbereicht 2015,Bieneninstitut Kirchhain, Kirchhain, Germany.

Uzunov A., Büchler R. (2018) *Results from brood interruption studies in Kirchhain.* COLOSS Varroa Control Task Force Workshop, 27-28 February 2018, Zadar, Croatia.

Uzunov A., Büchler R. (2020) *Alternative methods for varroa control in honey bee colonies.* Farmahem, Skopje, Nature Conservation Programme in North Macedonia, Swiss Agency for Development and Cooperation – SDC.

Veto-pharma (2015) *Varroa.* Veto-pharma; New York, USA.

Wilfert L., Long G., Leggett H.C., Schmid-Hempel P., Butlin R., Martin S.J.M., Boots M. (2016) Deformed wing virus is a recent global epidemic in honey bees driven by varroa mites. *Science*, 351: 594-597. https://doi.org/10.1126/science.aac9976

Winston M. L. (1987) *The biology of the honey bee.* Harvard University Press, Cambridge, Massachusetts, USA / London, UK. ISBN: 9780674074095

ABOUT THE AUTHORS

Dr Aleksandar Uzunov started keeping bees in 1995 when he was a student, and completed his PhD in 2013 on the diversity and biology of the Macedonian bee *Apis mellifera macedonica*. From 2015 to 2021 he was an external expert at the Bee Institute in Kirchhain, Germany. He is currently an Associate Professor and Head of the Laboratory for Honey Bee Biology and Breeding at the Ss. Cyril and Methodius University in Skopje, Macedonia, and a consultant at the Bee Institute (CAAS), Beijing, China.

Martin Gabel started beekeeping at the age of 15 and studied biology with a focus on ecology and nature conservation. Currently, Martin is working as Deputy Head of the Kirchhain Bee Institute, where he is involved in projects on bee pollinator diversity and forestry, as well as on his doctoral studies which are aimed at a better understanding of resistance mechanisms as applied to honey bee breeding and varroa research.

Dr Ralph Büchler has been a beekeeper for more than 50 years, and over his career has researched honey bee pathology, genetics and honey bee breeding, colony management schemes, and the effects of modern agriculture on the bee industry. Until 2022 he was Head of the Kirchhain Bee Institute, has given numerous lectures at national and international meetings of scientists and beekeepers, and written many papers in scientific journals, technical reports and book contributions.

The authors all share the philosophy of keeping bees naturally without causing adverse side effects on the environment, bees or consumers. Under diverse environmental and beekeeping conditions, Alex, Martin and Ralph have explored, studied and promoted beekeeping practices to produce safe hive products by using a combination of natural resistance and selection mechanisms. What Ralph started some twenty years ago as applied research in biotechnical methods for varroa control followed in the footsteps of his predecessors at the Kirchhain Bee Institute, has been endorsed internationally, and grown with Alex's support. A few years later, Martin, in the early days of his studies, also recognized the potential of nature-based beekeeping. He joined the team in 2015 and provided a new impetus to research and develop practices and techniques to meet beekeepers' expectations.

The team have since initiated numerous studies and published more than 50 articles, abstracts, communications and reports about biotechnical methods for varroa control, mainly focused on brood interruption methodologies. They are still working on the topic today, mainly within the international framework of the COLOSS Varroa Control and Survivors task forces. Together with colleagues from around the globe, approaches for varroa control continue to be developed, with studies on both the biological background and their applicability in practice, providing a new horizon for modern nature-based beekeeping practice.

www.ingramcontent.com/pod-product-compliance
Lightning Source LLC
Chambersburg PA
CBHW041309110526
44590CB00028B/4301